THE SHIELD AND THE SEED

Peace and Hope

THE SHIELD AND THE SEED

Peace and Hope

Mahendra Jagir

Studio of Books LLC
5900 Balcones Drive Suite 100
Austin, Texas 78731
www.studioofbooks.org
Hotline: (254) 800-1183

Ordering Information:
Special discounts are available on quantity purchases by corporations, associations, and others. For details, contact the publisher at the address above.

Printed in the United States of America.

ISBN-13:	Paperback	978-1-970283-42-6
	Hardback	978-1-970283-51-8
	eBook	978-1-970283-43-3

Library of Congress Control Number: 2026907052

DEDICATION

To my beloved grandchildren—Dustin, Matthew, Georgia, and Jolie.

You are the reason these words were written and the inspiration behind every page of these four books. In many ways, you are the "Silent Seeds" of my own life—the living promise of tomorrow and the harvest I am most proud to witness. Every idea I have explored, every system I have designed, and every story I have written has been guided by the hope that your generation will inherit a safer, wiser, and more peaceful world.

Throughout my life, I have traveled between two very different realms: the vast, open sky and the quiet, enduring strength of the earth. As a pilot, I spent countless hours above oceans and mountains, navigating the delicate balance between human ingenuity and the powerful forces of nature. As an engineer and planner, I learned to look beneath the surface of the land, to study its hidden structures and imagine new ways to protect and sustain the life that grows upon it. These experiences taught me that the future of humanity depends not only on our ability to dream, but also on our willingness to design systems that safeguard the world for generations yet to come.

Everything I have done—every mile I flew across the Pacific, every tunnel I imagined beneath the ground, and every protective system I envisioned standing guard against fire and climate—was done with your faces in my mind. You were the quiet motivation behind the long hours of reflection and the careful shaping of ideas that eventually became these books. In many ways, this series is not simply a collection of writings, but a personal legacy meant to guide, inspire, and protect.

I leave you these four books not merely as a library of thoughts, but as a map—one that reflects the journey of a lifetime and the lessons gathered along the way.

When you search for purpose and the courage to follow your dreams, read My Dream, My Destiny. Within its pages you will find the story of perseverance, the belief that even the most distant goals can be reached with determination and faith.

When you seek solutions for the challenges facing our world, read The SAN Project. It represents my belief that humanity can build systems that protect our planet while sustaining life for future generations.

When you desire wisdom, reflection, and a deeper sense of peace with nature, read The Silent Seeds. That book carries the quiet philosophy that true strength often grows slowly, like seeds beneath the soil.

And when you seek security and the confidence to stand strong in a changing world, read The Shield. It is my vision of a future where communities and homes are protected not through fear, but through thoughtful design and harmony with the natural forces around us.

I never had the opportunity to meet my own grandfather, yet I felt his presence through the land he worked and the legacy he left behind. His influence was carried not in spoken words but in the quiet strength of the soil beneath our family farmhouse. In the same way, I hope that through these pages you will feel my presence—not just as an author, but as a grandfather who cared deeply about the world you would one day inherit.

Consider these books as markers along a path I have tried to clear for you. I have set the posts, I have mapped the terrain, and I have built the shield as best I could with the knowledge and experiences of my lifetime. The surface world—the hills, the oceans, the forests, and the communities we build together—is yours to explore and enjoy.

Walk it with pride. Protect it with honor. And always remember that the greatest strength in this world is often the kind that remains unseen: the quiet systems that sustain life, the silent guardians that keep watch, and the enduring values passed from one generation to the next.

With all my love and faith in the future,

Mahendra Jagir

Table of Contents

PREFACE

THE COCKPIT OF PROTECTION

In my decades as a pilot, I learned a lesson that never left me: safety is never an accident—it is an architecture. When an aircraft is flying at 30,000 feet, there is no shoulder on which to pull over if something goes wrong. Every system, every sensor, and every safeguard must already exist within the machine before the journey begins. In aviation, survival depends on preparation long before the moment of crisis.

That same principle applies to the world we live in.

For many years I have watched California—my second home—transform into a landscape shaped by seasonal fear. Communities that once felt secure now face the growing unpredictability of climate extremes. Wildfires ignite with devastating speed, winds carry embers across valleys, and electrical grids fail at precisely the moment when stability is most needed. Beautiful homes, carefully built lives, and entire neighborhoods can suddenly find themselves vulnerable to what I often call the "Atmospheric Tantrum" of climate change, as well as the tragic unpredictability of human actions such as arson.

We have built our lives primarily on the surface of the earth, yet we have left those lives exposed. Our systems of protection often depend on distant fire stations, aging infrastructure, and emergency responses that arrive only after danger has already begun. In aviation, such a model would never be considered acceptable. Protection must be integrated into the design itself.

THE SHIELD was born from this realization. It is the fourth pillar of my life's work. If the Subterranean Arterial Network—the SAN Project—serves as the hidden artery of a resilient nation, then THE SHIELD functions as the immune system of the home. Together they represent a philosophy of protection that does not wait for disaster to occur. Instead, they prepare quietly and continuously, ensuring that stability already exists when the unexpected arrives.

Yet the ideas behind these systems reach beyond engineering alone. They emerge from a deeper belief about the nature of peace itself.

We live in an era where the word *peace* is spoken often but understood rarely. Too frequently, peace is treated as something fragile—a brief pause between conflicts, a silence between storms, a momentary absence of trouble. Nations celebrate it, communities long for it, and families pray for it. But too often peace is approached as something we hope for rather than something we build.

Hope is important, but hope alone is not enough.

Hope is a seed, and a seed must be planted in prepared soil. It must be protected, nourished, and given the structure necessary to grow. Without that protection, even the strongest seed can be lost to the winds of chaos. This book begins with a simple but powerful belief: peace should not exist merely as a feeling or a wish. Peace must become infrastructure.

Infrastructure endures where emotion cannot. Infrastructure continues its work quietly and faithfully regardless of circumstance. When peace becomes part of the systems that sustain life, it no longer depends on personalities, temporary agreements, or fortunate timing. It becomes dependable. It becomes resilient. It becomes real.

This understanding did not originate in a laboratory or engineering studio. It began much earlier in my life, during my childhood in Fiji.

Growing up in the highlands, I witnessed a wisdom that many modern societies have slowly forgotten. The elders did not speak about peace as an abstract concept. They practiced it through preparation. They respected the wind, the water, and the soil not as resources to be exploited, but as partners in survival. They understood that strength alone could never guarantee safety. True security required foresight.

Windmills turning slowly against the horizon were not merely machines.

They represented continuity and stability. The underground water channels carved into the earth were not simply irrigation systems. They were lifelines that ensured nourishment would reach the land when it was needed most. Seeds placed carefully into the soil were more than crops for the coming season. They were promises to the future—quiet declarations that life would continue.

These were not instruments of conflict. They were systems of care.

Over time, as the world accelerated and technology advanced, many of these lessons faded from collective memory. Human ingenuity increasingly focused on responding to threats rather than preventing them. Brilliant minds devoted their energy to the development of weapons, defensive barriers, and systems designed to overpower adversaries rather than sustain communities. Gradually, society began to believe that protection required domination, and that security could only be achieved through force.

Yet this belief contains a dangerous flaw.

Peace built on fear is never stable. Peace built on control is never complete. Peace built on destruction inevitably carries the seeds of future destruction. Such systems create cycles rather than solutions. They postpone conflict rather than preventing it.

True peace must be designed differently.

True peace protects without threatening, sustains without exhausting, and endures without constant defense. It must exist not as a fragile agreement but as a reliable system woven into the structure of everyday life.

This realization eventually led to the creation of two interconnected ideas: the SHIELD and the SAN.

The SHIELD Project began with a simple question: what if protection could stand silently, like a watchful elder—never harming, yet always prepared? What if structures designed to generate renewable energy could also serve as guardians of stability? What if the wind itself could become a partner in preservation?

The SHIELD windmill sentinels embody this philosophy. They stand tall not as instruments of war but as symbols of vigilance. They do not threaten or attack. Instead, they observe, endure, and quietly protect the communities around them. Their presence reminds us that the most effective guardian is not the one that reacts to destruction, but the one that prevents destruction from ever taking root.

Beneath the surface, the SAN—the Subterranean Arterial Network—serves as the unseen companion to the SHIELD. While the SHIELD stands above ground, visible and steadfast, the SAN flows quietly below, safeguarding water, energy, and essential resources. It ensures continuity even when uncertainty arises above the surface. Stability is achieved not through force, but through thoughtful design and long-term preparation.

Together, these systems form a comprehensive architecture of protection—one built not on violence, but on wisdom.

Yet beyond their engineering, the SHIELD and the SAN represent something deeply personal.

They are, in many ways, a manifestation of responsibility. The responsibility that each generation carries when it considers the world that will be inherited by the next. Every beam, every underground channel, and every silent mechanism within these systems reflects a promise: that the future deserves preparation, and that those who come after us deserve stability.

This book is therefore more than a technical proposal. It is an invitation to think differently about peace, resilience, and protection. It asks readers—young and old—to recognize that lasting security does not begin with fear or reaction. It begins with thoughtful planning and patient care.

Children, in particular, hold a remarkable capacity for this kind of thinking. They see the world not only as it is, but as it might become. They are not yet confined by the assumption that problems must be solved the same way they have always been solved. They are the future engineers, builders, and caretakers of our shared planet.

To them, this book offers both a vision and an invitation. Every great structure begins as an idea. An idea, carefully nurtured, becomes a plan. A plan, thoughtfully executed, becomes a structure. And a structure, maintained with care, becomes a legacy.

The SHIELD and the SAN are not final answers. They are the beginning of a new philosophy—one that values foresight over reaction, preparation over panic, and stewardship over conflict.

This is the vision of **Shatterproof Peace**: a peace that does not fracture under pressure, a peace that does not depend on fear, and a peace that stands quietly generation after generation.

And now, this blueprint is placed into your hands.

INTRODUCTION

THE SILENT SENTINEL

The philosophy of this book can be expressed in a single principle: self-reliance.

We are entering an era in which individuals and communities must be empowered to protect their own "Sovereign Soil." The systems that sustain modern life—energy, water, and safety—can no longer depend entirely on distant infrastructure or delayed response. True resilience must exist where people live. It must be present in the homes, communities, and landscapes that shape everyday life.

Drawing from the same principles I once applied as a private pilot in California, USA, and from the lessons I learned growing up in the highlands of Fiji, I began to imagine a different approach to protection. The answer, I realized, was not hidden in distant technologies or complex new resources. The answer already surrounds us. It exists in the wind moving above the land, the water resting beneath the soil, and the energy stored quietly within the earth itself.

This book is not merely about sprinklers, windmills, or mechanical systems. At its heart, it is about something far more valuable: peace of mind. It is written so that future generations—especially my grandchildren, Dustin, Matthew, Georgia, and Jolie—may grow up in the beauty of the hills without looking toward the horizon with unease. They deserve a world where the landscape inspires confidence rather than dread.

To understand why such systems are necessary, we must first reflect on the path humanity has followed.

For much of history, civilization measured its progress by its ability to overcome threats. Entire societies organized themselves around the concept of protection, and protection was most often defined by strength. The strongest walls promised safety. The most powerful weapons promised survival. The ability to defend, resist, and overpower became symbols of intelligence and preparedness.

The brightest minds of every generation accepted the responsibility of ensuring that their communities would endure whatever dangers the future might bring. With determination and creativity, they developed materials into shields, shaped stone and steel into fortresses, and designed systems capable of detecting and responding to danger. Their work reflected genuine dedication to preserving life.

Yet beneath this well-intentioned effort existed an important limitation.

Humanity became highly skilled at responding to danger, but far less experienced at preventing danger from emerging in the first place. It learned how to resist instability, but not always how to stabilize the systems that sustain life. It learned how to defend against destruction, but not how to build environments where destruction would struggle to take root.

Destruction, no matter how precise or technologically advanced, can only solve the problem of a single moment. It may stop an immediate threat, but it cannot ensure that new threats will not arise tomorrow. It can create temporary security, but it cannot establish lasting peace.

Destruction is reactive. It waits for instability before it acts.

Construction, however, offers something entirely different.

Construction prepares before instability appears. It builds systems that function quietly and continuously, creating stability within the foundations of everyday life. It establishes structures that endure, reducing the need for constant defense because resilience has already been designed into the environment itself.

This realization represents an important turning point in human understanding. The future will not be secured solely by the ability to eliminate threats. Instead, it will depend on the ability to construct systems that protect life steadily, patiently, and sustainably.

This shift in thinking arrives at a moment when it is urgently needed.

Across the world, people are witnessing environmental changes that cannot be ignored. Summers stretch longer than they once did, bringing heat that tests the limits of comfort and safety. Storms gather with unfamiliar intensity, reminding communities of the immense forces present in nature. Water arrives unpredictably—sometimes in overwhelming abundance and sometimes in troubling scarcity.

These developments are not isolated events. They are signals that the systems supporting human life must evolve.

The Earth continues to operate according to balance. It provides air to breathe, water to drink, soil to cultivate, and energy to sustain movement and growth. Yet balance requires cooperation. When equilibrium is disturbed, the Earth responds in order to restore it. These responses are not acts of hostility; they are natural corrections.

Humanity now faces the responsibility of adapting alongside the planet that sustains it. This adaptation demands more than temporary solutions. It requires a transformation in how protection itself is understood.

Protection can no longer depend solely on reaction. It must be embedded within the structures that support everyday life.

This understanding became the foundation of the SHIELD Project.

The SHIELD Project emerged from a simple yet powerful question: *What if protection could exist without destruction?* What if stability could be achieved through partnership with the natural world rather than through opposition to it?

At first glance, the SHIELD appears modest. Rising from the land like a windmill, its blades turn steadily with the currents of the wind. Its motion is calm, continuous, and purposeful. It neither rushes nor rests. It moves in harmony with forces that have existed long before humanity and will continue long after.

Yet its quiet presence carries remarkable significance.

The SHIELD transforms wind into energy and energy into stability. That stability supports homes, schools, hospitals, and the countless systems that allow communities to function. It provides power without exhausting irreplaceable resources. It strengthens infrastructure without creating harm.

More importantly, the SHIELD represents a new understanding of strength. It demonstrates that protection does not require aggression. Stability can emerge from cooperation rather than domination. Nature's forces need not be enemies to overcome—they can become partners in preservation.

Where the SHIELD stands, communities gain more than energy. They gain confidence. They gain reassurance. They gain the knowledge that their future rests upon systems designed with endurance in mind.

But stability cannot rely on a single structure alone. Lasting resilience requires networks—systems that support one another under all conditions.

This is where the Subterranean Assurance Network, or SAN, fulfills its essential role.

The SAN operates quietly beneath the surface of the earth. It carries water, energy, and vital resources through protected underground pathways, shielding them from disruption while ensuring continuity for the communities above. Like the roots of a great tree, the SAN remains largely unseen, yet its presence is indispensable.

Without roots, a tree cannot stand. Without underground support, stability on the surface cannot endure.

The SAN ensures that water continues to flow where it is needed. It safeguards energy distribution. It preserves the fundamental resources that sustain civilization even when surface conditions change.

Together, the SHIELD and the SAN create a unified system of resilience. One gathers strength from the wind above the earth, while the other preserves life beneath it. Their partnership creates balance. Their cooperation ensures continuity.

Together, they form what may be called Shatterproof Peace.

Shatterproof Peace does not depend on perfect conditions or temporary agreements. It exists because it has been carefully designed to endure uncertainty. By protecting the foundations of survival—water, energy, and food—it allows communities to continue living, growing, and thriving regardless of external challenges.

This vision belongs to every part of the world. It belongs to small islands surrounded by ocean, where the relationship between nature and survival has always been understood. It belongs to vast continents where millions rely on systems that must function reliably every day. It belongs to every community, every family, and every child.

Every child deserves to grow up in a world where the foundations of life are secure.

The responsibility of each generation is to build the systems that will protect the next. The builders of yesterday created the world we inhabit today. The builders of today will shape the world of tomorrow.

Humanity now stands at a moment of transition—from destruction to construction.

True strength lies not in the ability to destroy what threatens us, but in the wisdom to build systems that sustain life. Peace is not something that appears by chance. It must be designed, constructed, and maintained with care.

The SHIELD and the SAN represent the beginning of that effort. They symbolize a future in which protection is achieved through preparation rather than fear, and where stability is created through thoughtful design.

This book exists to guide that vision. It offers a blueprint for a world in which peace is engineered, stability is sustained, and the future is protected.

Because the future will not be shaped by destruction.

The future will be shaped by construction.

And that construction begins now.

CHAPTER 1

THE VULNERABILITY OF THE SURFACE

To understand the solution, we must first examine the vulnerabilities within the systems upon which modern society depends.

Across much of the developed world—and particularly in fire-prone regions such as California—our homes are constructed primarily as passive structures. They are designed to shield us from rain, wind, and cold, offering comfort and stability in ordinary conditions. Yet these same structures were never truly engineered to defend themselves against the rapidly intensifying environmental threats of the modern era. They provide shelter, but they do not actively protect themselves. Instead, they remain dependent on distant infrastructure and outside responders for safety.

This dependency becomes especially visible during wildfire events.

When fires ignite in vulnerable landscapes, one of the first preventive measures taken by authorities is often the deliberate shutdown of the electrical grid. This precaution, known in many regions as a public safety power shutoff, is intended to prevent damaged or sparking power lines from igniting additional fires during periods of extreme wind and dryness. While the measure can reduce the likelihood of new ignitions, it introduces an unintended vulnerability.

For many homes—particularly those in rural or hillside communities—electricity is not merely a convenience. It is the force that powers the pumps that draw water from wells, the systems that maintain pressure in storage tanks, and the infrastructure that supports emergency sprinklers or irrigation lines. When the electricity disappears, the water pressure often disappears with it. In a matter of moments, systems designed to support protection become silent.

At precisely the moment when defense is most necessary, the mechanisms that might have provided it fall dormant.

The home itself becomes exposed—not because its occupants failed to prepare, but because the structure was never designed to function independently of the larger grid. Pumps stop turning. Sensors go dark. Lighting fades. The technological lifelines that connect the household to the broader network vanish with the flip of a distant switch.

Meanwhile, the fire continues its advance.

Wildfires rarely appear suddenly at the edge of a neighborhood. Long before flames are visible, the surrounding environment begins to change. Radiant heat travels outward from the fire front, warming surfaces far beyond the immediate boundary of the flames. Structures absorb this heat gradually, and materials begin to respond.

Wood becomes increasingly dry. Window frames expand slightly under temperature stress. Roofing materials lose residual moisture. Vegetation surrounding the home becomes brittle and highly combustible. By the time smoke becomes visible on the horizon, the process that prepares a structure for ignition may already be underway.

In this way, wildfire operates as both a distant and an approaching threat. The damage does not begin when the flames arrive; it begins much earlier, as heat silently alters the environment.

Despite this reality, many of the systems we rely upon for protection remain largely reactive.

Someone must first notice the smoke rising above the hills. A call must be placed to emergency services. Dispatch centers must analyze the information and determine the appropriate response. Fire crews must assemble, vehicles must travel across miles of winding roads, and aircraft must be mobilized to survey and combat the growing blaze.

Each step is essential. Each step is necessary. Yet each step also requires time.

Fire, by contrast, requires none.

Driven by wind, terrain, and accumulated fuel, modern wildfires can move with extraordinary speed. They climb slopes rapidly, accelerate through canyons, and leap across natural barriers with embers carried far ahead of the main flame front. In many cases, the fire evolves faster than human response systems can mobilize.

The imbalance becomes impossible to ignore. The threat moves continuously, yet the response often begins only after the danger has already intensified.

This vulnerability is not simply technological—it is structural. For generations, communities were built with the expectation that protection would arrive from elsewhere. Fire stations, regional response teams, centralized power grids, and municipal water systems were assumed to be available whenever danger appeared.

Under normal circumstances, this model functions well. But when extreme conditions interrupt those networks—when roads close, power shuts down, or emergency services are stretched beyond capacity—the weaknesses in that assumption become painfully clear.

Entire neighborhoods can be left temporarily unprotected, not through negligence, but through the sheer scale and speed of the crisis.

It was from the vantage point of the sky that this pattern became most evident to me.

During my years as a pilot, I spent countless hours flying over landscapes that stretched across mountains, valleys, forests, and coastlines. From the ground, the movement of fire often appears chaotic and unpredictable. Yet from above, a different pattern emerges. Fires reveal a form of logic when observed from the air.

They follow the contours of the terrain. They race through narrow valleys where wind accelerates naturally. They climb slopes where rising heat intensifies combustion. They spread most rapidly through corridors of dry vegetation that have accumulated over decades.

In effect, the landscape itself becomes a map that guides the fire's movement.

In California particularly, the relationship between wind and wildfire is unmistakable. Seasonal winds sweeping inland from the coast can push flames across great distances. Even more dangerous are the embers carried high into the air and transported miles ahead of the fire front. These embers can ignite roofs, dry leaves in gutters, or vegetation within neighborhoods long before the main fire arrives.

Entire communities may ignite not because the flames reached them directly, but because the wind delivered sparks into their midst.

Watching these patterns unfold from the cockpit, a realization gradually took shape.

Wind is one of the fire's most powerful allies.

Yet wind is also one of the most abundant natural resources available to us.

Every day, currents of air move across mountains and plains without interruption, carrying vast amounts of kinetic energy. For decades, engineers have learned to capture this energy using wind turbines, converting the motion of the atmosphere into electricity that can be stored and used later.

The implication is profound.

If wind can spread the fire, it can also power the systems that defend against it.

This concept represents a shift in perspective. Instead of relying entirely on distant infrastructure or emergency responders, communities could integrate protective systems directly into the landscape itself. Energy gathered from the environment during calm and stable days could be stored locally and deployed during moments of crisis.

In such a system, protection would no longer depend solely on human reaction.

Most conventional safety systems require observation, decision, and action. Someone must notice the danger. Someone must determine the appropriate response. Someone must activate the equipment.

But wildfire often moves faster than this chain of decisions allows.

What is needed instead is a form of continuous vigilance— something that observes the environment at all times and reacts instantly when danger appears.

In other words, a sentinel.

A sentinel does not sleep. It does not hesitate. It observes patiently and acts the moment a threat emerges.

Modern technology now makes this concept achievable. Infrared sensors, thermal cameras, and environmental monitors are capable of detecting subtle increases in temperature or the earliest thermal signatures of ignition. Long before flames are visible to the human eye, these systems can recognize abnormal heat patterns developing in the surrounding landscape.

Detection can occur within seconds.

Yet detection alone is not enough. A warning without a response changes nothing. For protection to be meaningful, the system must also possess the ability to act immediately and effectively.

That requires stored energy, accessible water, and a distribution system capable of delivering that water precisely where it is needed.

This is where the concept of the Water Shield begins to emerge.

Water has always been one of humanity's most reliable tools in the fight against fire. Firefighters around the world depend on it because it cools surfaces, absorbs heat, and prevents combustible materials from reaching ignition temperatures.

Traditionally, however, water is deployed only after a fire has already arrived.

A Water Shield reverses this sequence entirely.

Rather than reacting to flames, the system prepares the structure in advance. Roofs, exterior walls, decks, and surrounding vegetation can be moistened before radiant heat reaches critical levels. The environment surrounding the structure becomes saturated, reducing the likelihood that embers will ignite vulnerable materials.

Moisture replaces dryness. Heat is absorbed rather than accumulated.

The objective is not necessarily to extinguish the wildfire itself—a task often beyond the capacity of any single home or community system. Instead, the goal is to create a protective envelope around the structure, dramatically reducing the probability of ignition.

Timing is the critical factor.

If water is applied too late, the structure may already be compromised. But when applied early—while the fire remains at a distance—the protective effect can be substantial. Surfaces remain cooler, combustible materials retain moisture, and embers are far less likely to ignite new flames.

When powered by energy harvested from wind and stored locally, such a system could operate independently of the electrical grid. Even during widespread power shutdowns, the sentinel would remain active. Sensors would continue to monitor the environment. Pumps would remain ready. Water would remain available.

No human command would be required.

The system would simply perform the task for which it was designed.

In this way, the wind—the same force capable of carrying embers across great distances—would also become the energy source that powers the defense against them.

The transformation is both practical and philosophical.

Rather than building passive homes that depend entirely on rescue, we can begin designing communities that participate actively in their own protection. Structures would no longer wait helplessly as danger approaches. Instead, they would respond intelligently, automatically, and immediately.

The vulnerabilities of the surface world have been visible for decades. Yet vulnerability does not have to remain permanent. Once we recognize the weaknesses in our current systems, we gain the opportunity to redesign them.

This realization leads to a broader philosophy.

Protection cannot depend on reaction alone. It must be woven directly into the fabric of the environment. It must combine ecological restoration, resilient infrastructure, and intelligent technological guardians working together as a unified system.

The Silent Seeds restore the stability of the natural world, strengthening soil health, preserving biodiversity, and ensuring that food systems remain sustainable.

The Subterranean Assurance Network—SAN—protects the circulation of water and energy beneath the surface, safeguarding the invisible lifelines that sustain modern civilization.

And the SHIELD rises above the landscape as a visible sentinel, harvesting wind energy and transforming it into continuous protection.

Together, these systems form an integrated architecture of resilience.

They represent a shift away from reactive defense toward proactive stability. They demonstrate that the forces of nature need not be feared or resisted. When properly understood, they can become allies in preservation.

This vision is not intended for the present alone. It is meant for the generations that will inherit the world we shape today.

To Dustin, Matthew, Georgia, Jolie, and to the many young people who will grow into leadership in the decades ahead: the challenge your generation will face is not a war between nations, but a struggle against instability itself. Climate disruption, resource strain, and environmental uncertainty will test the systems we have built.

The solution will not come from greater destruction.

It will come from greater wisdom in how we design the world around us.

Your role will not be to fight endless battles against nature. Your role will be to maintain, refine, and improve systems that work in harmony with it.

The ancient wisdom I witnessed in Fiji taught that preparation is the highest form of protection. Communities that respect the rhythms of the land survive storms that others cannot.

Modern engineering now provides the tools to apply that same wisdom on a global scale.

The ideas described in this chapter represent only the beginning of that effort. They are the first steps toward a civilization designed not merely to endure disaster, but to prevent it.

And the foundation of that vision begins with a simple but powerful idea:

A sentinel that never sleeps.

CHAPTER 2

THE SUBTERRANEAN RESERVOIR (THE SOURCE)

The first layer of THE SHIELD begins not in the sky, but deep within the earth itself. Long before technology rises above the landscape in the form of windmill sentinels, true protection must begin below the surface, where the natural stability of the planet quietly safeguards one of life's most essential resources: water. It was in this hidden world beneath our feet that the philosophy of the "Silent Seeds" first took root, teaching that the most powerful systems are often the ones that operate quietly, out of sight, yet with profound influence on the survival of everything above them.

In regions where environmental risk is high—especially in landscapes vulnerable to drought, wildfire, or infrastructure failure—every property must possess its own source of water that does not depend on distant networks or fragile surface systems. This concept is embodied in what I call the **Subterranean Sovereign Well**. Unlike municipal pipelines that stretch across miles of exposed terrain, the Sovereign Well reaches downward into the aquifers that lie protected beneath layers of soil and stone. These underground reservoirs have existed for thousands of years, shielded from the extreme temperature swings and destructive forces that shape the surface world.

Centralized water systems provide convenience under normal conditions, but crises reveal how vulnerable those systems can become. Pipelines may fracture under intense heat. Electrical grids that power pumping stations can be deliberately shut down to prevent additional fires. Water pressure may collapse when entire communities attempt to draw from the same network during emergencies. What once appeared dependable can suddenly disappear at the very moment when protection is needed most.

True resilience therefore begins with independence.

The Subterranean Sovereign Well ensures that independence by establishing a direct connection between the home and the deep aquifers below. These underground reserves remain naturally insulated by the earth itself. Even during extreme heat events or raging wildfires, the subsurface environment remains remarkably stable. While flames may consume vegetation above ground and temperatures may soar in the surrounding air, the soil and rock beneath the surface preserve water at consistent, cool temperatures.

Within the architecture of THE SHIELD, this well performs two vital roles. During ordinary life it functions much like the root system of a plant, supplying the daily "nutrients" that sustain the household. Water drawn from the aquifer supports drinking, sanitation, agriculture, and the broader ecological balance of the property. In this way the well quietly nourishes life day after day, much like the unseen roots that sustain a forest.

Yet during a crisis, its role changes dramatically.

When the SHIELD system enters emergency mode, the well becomes the primary defensive resource of the entire structure. The same water that nourishes life is transformed into a powerful protective force capable of stabilizing the environment around the home.

Water drawn from the well is directed into a reinforced subterranean reservoir constructed according to the principles of the Subterranean Arterial Network, or SAN. This underground chamber functions as a secure vault designed to store large volumes of water beneath the insulating protection of the surrounding earth. Unlike surface tanks that may heat rapidly under direct sunlight or become compromised during disasters, the subterranean reservoir remains protected from external extremes.

The earth itself acts as a natural shield.

Even when temperatures above ground climb to extreme levels—as they often do during intense wildfire events—the water stored underground remains cool and stable. In many locations, groundwater temperatures remain close to fifty-five degrees Fahrenheit year-round. This difference between the heat above and the coolness below becomes one of the most powerful advantages of the entire SHIELD system.

When the system activates, pumps powered by locally generated energy propel this cool subterranean water through high-pressure distribution lines that surround the structure. Spray heads positioned along rooftops, walls, and surrounding vegetation release controlled streams that saturate vulnerable surfaces.

The result is not merely a thin coating of moisture.

Instead, the sudden introduction of cool water into an overheated environment produces a rapid thermal shift. Surfaces that were absorbing radiant heat are instantly cooled. Air temperatures surrounding the structure begin to drop. Materials that might otherwise approach ignition temperatures are stabilized before they can reach the critical threshold required for combustion.

This phenomenon can be understood as a form of localized thermal disruption. Fire depends on sustained heat to spread from one material to another. By introducing a sudden cooling effect, the SHIELD interrupts that process. The environment surrounding the protected structure becomes far less favorable for ignition.

In essence, the system creates a defensive climate around the home.

Flames approaching the structure encounter cooler surfaces, higher humidity, and materials saturated with moisture. Embers that land on rooftops or nearby vegetation are less likely to ignite dry fuel. Radiant heat that might otherwise prepare the structure for combustion is continuously absorbed and dissipated by the cooling water.

Through this simple but powerful process, the subterranean reservoir transforms water into a dynamic barrier against environmental threats.

Yet the SHIELD Project extends far beyond the well and reservoir alone. While the underground system provides the essential resource for protection, another layer of infrastructure rises above the landscape to provide awareness, energy, and coordination.

These guardians are the windmill sentinels.

Across hills, coastlines, plains, and mountain ridges, these tall structures stand quietly against the horizon. At first glance they resemble the wind turbines that have become symbols of renewable energy around the world. But within the architecture of THE SHIELD, their role extends far beyond the generation of electricity.

Each windmill functions simultaneously as a power generator and as an environmental sentinel.

Through the rotation of their blades, these towers harvest the kinetic energy of the wind, converting a naturally occurring force into clean electrical power. That energy is stored and distributed to support the entire protective system—powering pumps, sensors, communication systems, and monitoring equipment that operate continuously around the property.

Because energy is generated locally rather than drawn entirely from distant power plants, the system becomes far more resilient. If regional electrical grids fail or are deliberately shut down during emergencies, the SHIELD continues operating independently. Energy gathered from the wind ensures that the infrastructure responsible for protection remains active even during widespread outages.

Yet energy generation is only one part of the windmill's purpose.

Each sentinel is equipped with advanced environmental sensors capable of monitoring subtle changes in the surrounding atmosphere. These instruments observe temperature fluctuations, humidity levels, wind speed, pressure patterns, and other environmental indicators that reveal the health of the landscape.

Through this continuous observation, the infrastructure itself becomes aware of its surroundings.

Instead of relying solely on human observers to recognize danger, the SHIELD system analyzes environmental data in real time. When unusual patterns begin to emerge—such as rapid temperature increases, abnormal thermal signatures, or atmospheric conditions associated with wildfire ignition—the system can identify these changes within moments.

In this way the windmill sentinels become the first line of awareness.

They do not wait for smoke to be visible from miles away. They detect the earliest signals of instability long before a crisis becomes obvious to the human eye. This early detection dramatically shortens the time between the appearance of a threat and the activation of protective measures.

Energy independence further strengthens the resilience of this design. Modern infrastructure often relies heavily on centralized power grids, yet those grids are among the first systems to fail during natural disasters. Storms, fires, mechanical breakdowns, and safety shutdowns can interrupt electricity across vast regions.

By decentralizing energy production through a distributed network of windmill sentinels, the SHIELD system avoids this vulnerability. Each tower contributes to the overall energy supply, creating redundancy throughout the network. If one tower becomes damaged or temporarily disabled, the remaining structures continue generating power.

The system therefore becomes flexible rather than fragile.

Another important feature of the windmill sentinels lies in their visibility. Much of modern infrastructure operates invisibly—hidden beneath streets, buried underground, or located far from the communities it serves. While these systems may function effectively, their hidden nature often leaves residents unaware of the mechanisms protecting them.

The sentinels change this dynamic.

Standing tall across the landscape, they provide a visible reminder that protective systems are already in place. Communities living near these structures can look toward the horizon and see physical evidence that the land is being monitored and safeguarded.

This visibility has a profound psychological effect.

When people know that protective systems are active and reliable, fear begins to diminish. Instead of feeling exposed to unpredictable disasters, communities gain confidence that the environment around them has been designed with resilience in mind.

In this way the windmill sentinels perform both mechanical and social roles. They generate energy, monitor environmental conditions, and strengthen the collective sense of security within the communities they serve.

Perhaps the most significant aspect of the SHIELD Project lies in its commitment to prevention rather than reaction.

Traditional emergency systems often operate only after a crisis has already begun. A wildfire ignites, alarms are raised, and response teams mobilize as quickly as possible. Although these efforts are heroic and essential, precious time is often lost during the earliest stages of a threat—the moment when intervention might be most effective.

The SHIELD system changes this sequence entirely.

Because its sensors and monitoring systems operate continuously, the network can recognize subtle warning signals before a crisis escalates. Minor thermal anomalies, unusual atmospheric conditions, or early ignition signatures can trigger automated protective responses designed to stabilize the environment immediately.

Protective sprays may activate to increase moisture levels. Cooling systems may begin saturating vulnerable surfaces. Environmental adjustments may be initiated to reduce ignition risk.

By responding at the earliest possible stage, the system dramatically lowers the likelihood that a small spark will evolve into a devastating wildfire.

Instead of racing to extinguish a fully developed disaster, the SHIELD Project works quietly to prevent many disasters from forming in the first place.

At its heart, the SHIELD represents a transformation in the philosophy of protection. For centuries humanity has devoted immense resources to systems designed for conflict. Defensive technologies were created to defeat enemies, often producing destruction that harmed the very world they were meant to protect.

The SHIELD Project offers a different vision.

Its systems are not built to intimidate or destroy. They are designed to stabilize ecosystems and prevent the conditions that allow crises to grow.

Water replaces weapons.

Cooling replaces combustion.

Preparation replaces panic.

By strengthening environmental balance and preserving natural resources, the SHIELD infrastructure reduces the pressures that often lead to ecological disaster and resource scarcity. In this sense, infrastructure itself becomes a quiet instrument of peace.

The windmill sentinels rising along the horizon symbolize this transformation. They stand not as monuments to war, but as guardians of stability. Their blades harvest the wind, their sensors observe the land, and their energy sustains the systems that protect communities from environmental threats.

As the SHIELD network expands across vulnerable landscapes, energy becomes more reliable, ecosystems grow stronger, and the risks of catastrophic wildfire begin to diminish.

This is the promise of the SHIELD Project.

A future where protection is intelligent, sustainable, and peaceful.

A future where the guardians on the horizon watch silently over the land—not as instruments of conflict, but as instruments of balance and stability.

CHAPTER 3

THE SUBTERRANEAN RESERVOIR (THE SOURCE)

In aviation, redundancy is the foundation of safety. Pilots are trained to assume that even the most reliable systems may one day fail. For this reason, aircraft are designed with layers of backup mechanisms that ensure essential functions continue operating even during emergencies. One of the most important of these systems is the Auxiliary Power Unit, commonly known as the APU. If the main engines fail, the APU provides the lifeblood of electricity that keeps the aircraft's critical systems running. It powers the instruments, maintains communication, and sustains the aircraft long enough for the pilot to guide it safely to the ground.

The individualized windmill on your property serves the same purpose.

Within the architecture of the SHIELD Project, this residential-scale windmill functions as the energy engine of the entire protection system. It is not a massive industrial turbine dominating a distant ridgeline, but rather a refined and elegant structure designed specifically for individual properties. These windmills operate quietly and efficiently, capturing the natural movement of the wind and converting it into usable energy for the home and its surrounding protective infrastructure.

Paired with solar arrays, the windmill forms what I describe as a Hybrid Energy Seed. Throughout the year, this system silently harvests the two most reliable renewable resources available to us: the wind above and the sunlight that reaches every open landscape. Together, these sources provide a steady stream of energy that can be stored and used whenever the system requires it.

Energy storage is just as important as energy generation. In the SHIELD design, the electricity produced by the windmill and solar panels is stored within fire-hardened battery banks located underground. This subterranean placement is not accidental; it follows the same guiding principle that defines the Subterranean Arterial Network. The earth itself becomes a natural shield, protecting the most essential components of the system from extreme heat, wildfire exposure, and environmental damage.

By storing the system's energy underground, we ensure that even if exterior wiring or above-ground components are damaged by fire or severe weather, the heart of the system continues beating safely below the surface. The batteries remain insulated, stable, and ready to deliver power when it is needed most.

When the sky turns dark with smoke and the electrical grid collapses—as it often does during major wildfire events—the Hybrid Energy Seed becomes the property's independent power source. The stored energy activates the high-pressure pumps that drive the SHIELD's water distribution system, allowing the Subterranean Reservoir described in the previous chapter to release its cooling defense around the structure.

In this way, the individualized windmill does far more than generate electricity. It becomes the engine that drives the entire protective system. It provides the independence necessary to maintain safety even when external infrastructure fails.

Yet the philosophy behind the SHIELD Project extends beyond the protection of individual homes. It is rooted in a broader understanding of how modern civilization functions—and where its greatest vulnerabilities lie.

For generations, humanity has built its cities, roads, and infrastructure primarily on the surface of the earth. Power lines stretch across landscapes, pipelines travel above or just beneath the soil, and communication towers rise high into the sky. These surface systems allowed societies to expand rapidly and connect communities across great distances.

But the surface of the planet is also the most vulnerable place to build critical infrastructure.

Storms, floods, earthquakes, and wildfires strike the surface first. Power lines can collapse in strong winds. Transmission towers may fall during earthquakes. Pipelines can rupture under extreme conditions, and transportation networks can be disrupted in a matter of hours during major disasters. When these systems fail, the consequences ripple quickly through society. Electricity disappears, water pumps stop working, communications break down, and supply chains stall.

Modern civilization depends on the continuous movement of resources. When that movement is interrupted, even briefly, the stability of entire communities can weaken.

For a long time, this vulnerability was accepted as unavoidable. The surface was simply the easiest place to build. Infrastructure could expand rapidly, linking cities and regions together. But as environmental risks increase and populations grow, the limitations of surface-based systems have become increasingly clear.

If civilization is to become truly resilient, its most essential systems cannot remain exposed to constant disruption.

They must be protected.

This realization led to the development of the Subterranean Arterial Network, or SAN—a concept that applies one of nature's most powerful lessons to human infrastructure.

Beneath the surface of the earth lies an environment that remains remarkably stable. While storms rage above and temperatures fluctuate dramatically, underground conditions remain cool, steady, and insulated from the chaos of the atmosphere. Plant roots, underground water systems, and geological formations have long relied on this stability. Nature itself uses the earth as a protective shield for its most vital processes.

The SAN applies this same principle to the systems that sustain modern civilization.

Instead of placing critical infrastructure directly on the surface where it remains exposed to environmental threats, the SAN moves these systems underground. Pipelines, energy conduits, water channels, and communication networks are placed beneath layers of soil and rock, where they are naturally shielded from fire, wind, and extreme weather.

When wildfires sweep across forests, the underground network remains untouched. When storms destroy power lines above ground, energy continues flowing through protected conduits below. Even when surface transportation routes are blocked, the subterranean infrastructure continues delivering essential resources to the communities that depend upon them.

For this reason, the SAN can be thought of as the arterial system of civilization.

Just as arteries carry oxygen and nutrients throughout the human body, the Subterranean Arterial Network carries the essential lifelines that sustain modern society. Through this underground network flow water, energy, communication, and agricultural support systems— each moving continuously through protected channels that remain shielded from disruption.

Clean water travels through underground pipelines, ensuring reliable supply for homes, agriculture, and emergency services. Energy generated by wind and solar systems flows through secure conduits that reduce the risk of outages caused by storms or fires. Agricultural systems receive irrigation support from protected reservoirs, maintaining food production even during environmental stress. Communication networks and monitoring systems operate through shielded channels, ensuring that emergency services remain active during crises.

Together, these interconnected systems form a protected delivery network that allows resources to continue moving even when the surface environment becomes unstable.

This ability to maintain continuous flow is one of the SAN's greatest strengths. Modern societies rely on constant movement—electricity flowing through grids every second, water circulating through treatment plants and pipelines, and food arriving daily to markets and homes. Any interruption to this flow can create immediate and serious consequences.

The SAN is designed specifically to prevent those interruptions.

Because the network connects multiple sources of supply and operates through redundant pathways, it can automatically reroute resources if one section becomes unavailable. If damage occurs in one area, alternate channels continue delivering water and energy to the communities that need them.

In times of emergency, the system can also prioritize critical facilities such as hospitals, emergency shelters, and firefighting infrastructure. Water can be redirected to regions facing wildfire threats or drought conditions, while energy can be concentrated where it is most urgently required.

In this way, the network behaves less like a rigid machine and more like a living system—adapting continuously to maintain stability.

Every civilization rests upon a foundation. In the past, that foundation has often been fragile because it depended heavily on exposed systems that could fail under pressure. Power outages, water shortages, and supply chain disruptions revealed how quickly stability could be lost.

The SAN changes this equation.

By placing essential infrastructure underground, the network creates a protected layer that supports everything built above it. Cities, homes, farms, and renewable energy systems all rely upon this hidden foundation to maintain stability and continuity.

Unlike the towering windmill sentinels of the SHIELD Project, much of the SAN remains invisible to those living above it. Yet its quiet presence is precisely what allows the visible world to function. Beneath the soil, beneath the roads and buildings, the network continues its silent work—moving water, carrying energy, supporting agriculture, and maintaining the essential rhythms of modern life.

When the SHIELD Project and the Subterranean Arterial Network operate together, a new kind of infrastructure begins to emerge.

Above the ground, windmill sentinels harvest renewable energy and observe the landscape, serving as visible guardians that monitor environmental conditions. Below the surface, the SAN distributes that energy and the life-sustaining resources that communities depend upon.

One system watches the horizon.

The other strengthens the foundation beneath our feet.

Together, they form an integrated architecture of resilience—an infrastructure designed not merely to support civilization today, but to protect it for generations to come.

The SAN may remain unseen, but its influence is profound. It is the hidden strength of the future, the underground network that ensures life above the surface can continue even when the world becomes uncertain.

CHAPTER 4

THE INTEGRATED SENTINEL (SPRINKLER LOGIC AND THE KILL-SWITCH)

At the center of the SHIELD system lies its most important capability: the ability to transform instantly from a producer of energy into a guardian of protection. Under normal circumstances, the windmill functions as a quiet and dependable energy generator. Its blades rotate steadily in the breeze, harvesting the invisible motion of air and converting it into electricity that powers homes and strengthens the resilience of the surrounding property. Working alongside solar arrays and subterranean battery systems, the windmill forms part of what I describe as a Hybrid Energy Seed—a self-sustaining system that gathers energy from the natural forces already present in the environment.

Yet the same wind that makes this system productive during ordinary conditions can become dangerous during a wildfire emergency. Wind is the primary force that spreads fire across landscapes. It carries burning embers across forests, valleys, and neighborhoods, sometimes transporting sparks miles away from the original fire front. In such conditions, the motion of a rotating turbine can unintentionally contribute to the turbulence that allows embers to circulate around structures. What was once a peaceful energy generator can quickly become an element of vulnerability if it continues to operate during a crisis.

For this reason, the SHIELD system incorporates what I refer to as an Integrated Sentinel Logic. This design philosophy ensures that when a wildfire threat is detected, the windmill immediately ceases to function as an energy generator and instead becomes a critical component of the defensive system protecting the property. The transition is designed to occur automatically and within seconds, removing the need for human intervention during moments when every second may matter.

High above the property, mounted at the upper sections of the windmill tower, are large-scale suppression sprinklers engineered specifically for emergency environmental defense. These devices bear little resemblance to the small oscillating sprinklers commonly used for lawns and gardens. Instead, they are industrial-grade suppression heads designed to deliver extremely large volumes of water across wide areas with precision and force. Their purpose is not irrigation but rapid environmental cooling and fire suppression.

These suppression heads are connected through reinforced steel piping to the subterranean reservoir system described in earlier chapters. That reservoir draws water from deep aquifers located far beneath the surface of the earth. These underground water sources remain naturally insulated by layers of soil and rock, maintaining a stable temperature even when extreme heat affects the surface environment. While wildfires may drive surface temperatures to extraordinary levels, the water stored underground remains cool and steady, typically close to the natural temperature of deep groundwater.

When the defensive system activates, this cold water becomes one of the most powerful tools in protecting the structure.

The sequence begins with environmental detection. Sensors placed throughout the property monitor atmospheric conditions continuously. These sensors are calibrated with careful precision so that they can distinguish between the normal heat of a summer afternoon and the dangerous temperature patterns associated with approaching wildfire conditions. Infrared detection systems observe heat signatures across the landscape, while temperature monitors and atmospheric sensors track sudden fluctuations that may signal the presence of fire nearby. These instruments work quietly in the background, gathering information and comparing real-time environmental data against known wildfire patterns.

When the system detects a pattern that suggests a wildfire threat, it immediately initiates the next stage of defense.

The windmill receives a command signal that triggers an electromagnetic braking system within the turbine mechanism. This braking system brings the rotating blades to a controlled halt. Within moments, the windmill stops turning completely. In doing so, the structure transitions from a rotating turbine into a stationary observation tower.

This shutdown serves an important purpose. By stopping the blades, the tower eliminates the air turbulence that rotating blades could generate during high winds. Without this turbulence, embers carried by wildfire winds are less likely to be drawn toward the structure or circulated unpredictably around the property. The tower becomes stable and calm, ready to serve as a platform for the defensive systems that follow.

At the same time, the stored energy gathered throughout the year by the Hybrid Energy Seed is redirected toward emergency systems. Electricity stored in subterranean battery banks is released and routed directly to high-pressure pumps connected to the underground water reservoir. These pumps are engineered to generate powerful water pressure capable of delivering large volumes of groundwater through the suppression network.

As the pumps reach full power, the final stage of the system begins.

The large suppression sprinklers activate simultaneously, releasing powerful arcs of water across the roof of the structure and the surrounding perimeter of the property. Because the windmill blades have already stopped turning, the water disperses smoothly and evenly without interference from rotating air currents. Instead of scattering unpredictably, the spray pattern forms a consistent layer of cooling water across the building and its immediate surroundings.

Within seconds, the property becomes enveloped in what can best be described as a protective water dome.

This water dome performs several important functions simultaneously. First, it cools the surfaces of the structure, reducing the risk that roofing materials, walls, or nearby vegetation will reach ignition temperatures. Second, the introduction of large volumes of cold groundwater into an extremely hot environment produces a rapid drop in localized air temperature. This phenomenon creates a form of thermal shock that helps suppress the conditions necessary for combustion. Finally, any embers carried by the wind that land within the saturated zone are quickly extinguished before they can ignite surrounding materials.

In this state, the windmill tower serves a new purpose. No longer acting as an energy turbine, it becomes a sentinel standing above the property. From its elevated position, it allows water to rain down across the structure and surrounding area, extinguishing small ignition points before they can develop into larger fires.

Through this process, the SHIELD system buys critical time. Even if a wildfire passes nearby, the structure remains cooled, saturated, and far less vulnerable to ignition. What might otherwise have been a moment of catastrophe becomes a controlled defensive response guided by engineering and preparation.

Yet even a system as advanced as the SHIELD represents only one part of a much larger vision.

Protection against immediate threats is essential, but the long-term stability of human civilization depends on something more fundamental. The true foundation of lasting security lies not only in technological defense systems but also in the health of the environment itself. This is where the philosophy of Silent Seeds enters the framework.

Throughout history, civilizations have often measured their strength through visible achievements such as cities, monuments, machines, and expanding infrastructure. Yet beneath every successful civilization lies a quieter and more essential foundation: the condition of the soil, forests, and water systems that sustain human life. When societies maintain balance with the land that feeds them, they prosper. When that balance deteriorates, instability soon follows.

Silent Seeds represent a philosophy of restoration rooted in this understanding. These seeds symbolize more than agricultural development; they represent the rebuilding of the natural systems that support life. Instead of treating the environment as something to dominate or exhaust, Silent Seeds encourage cooperation with the natural processes that allow ecosystems to regenerate and sustain themselves.

In the modern agricultural world, food production has often been driven by speed and efficiency. Crops are frequently selected for rapid growth, uniform appearance, and high short-term yields. Large monoculture fields can produce impressive harvests under ideal conditions, but they also create fragile systems that depend heavily on fertilizers, pesticides, and precise environmental control.

Silent Seeds represent a different approach. They are selected and cultivated for resilience rather than perfection. These plants are capable of surviving heat, drought, shifting weather patterns, and less-than-perfect soil conditions. Their purpose is not simply to produce food but to help restore the health of the land itself.

As these resilient plants grow, their roots strengthen soil structure and help prevent erosion. They improve the soil's ability to retain water, allowing rainfall to soak into the ground rather than washing nutrients away. Over time, landscapes that were once degraded can begin to recover, gradually rebuilding fertility and ecological stability.

In this way, agriculture becomes more than food production. It becomes environmental restoration.

The principles behind Silent Seeds are not entirely new. Many traditional societies developed agricultural systems that reflected deep understanding of their local ecosystems. These systems were refined over generations through observation, experience, and careful stewardship of the land.

Among the landscapes that illustrate this knowledge are the islands of Fiji. The communities that lived there historically developed agricultural practices that balanced productivity with environmental sustainability. They recognized the importance of preserving forests to protect watersheds. They understood that soil must be allowed time to recover between planting cycles. They cultivated diverse crops that strengthened resilience against environmental change.

These practices were not written in technical manuals but passed down through cultural traditions and community knowledge. As the modern world expanded, many of these practices faded into obscurity. Yet their underlying wisdom remains remarkably relevant to the challenges facing the modern planet.

Silent Seeds therefore represent a bridge between ancient ecological understanding and modern technological capability. They combine the insights of traditional environmental stewardship with contemporary research and agricultural science.

Within the broader framework presented in this book, Silent Seeds form the environmental pillar that supports the entire system.

The SHIELD Project protects communities from immediate environmental threats such as wildfire.

The Subterranean Arterial Network ensures the reliable flow of water, energy, and essential resources beneath the surface of the earth.

Silent Seeds restore the living ecosystems that make both of those systems meaningful.

Together, these three elements create a balanced structure for long-term resilience. Protection alone is not enough. Infrastructure alone is not enough. Environmental restoration alone is not enough. But when all three operate together, they create a system capable of supporting human civilization in a changing and uncertain world.

Through this integration of protection, infrastructure, and environmental renewal, humanity gains something far more valuable than temporary security.

It gains resilience—the quiet foundation upon which lasting peace is built.

CHAPTER 5

THE ARSONIST AND THE ATMOSPHERIC TANTRUM

One difficult reality must be acknowledged when discussing fire: not all fires are natural. Some begin with lightning strikes, others with failing transformers or neglected power lines, and some with deliberate human actions. In an age of rising temperatures and prolonged drought, even a single spark—whether accidental or intentional—can trigger a disaster that spreads faster than communities can respond. The atmosphere itself has become increasingly volatile, turning seasonal heat into what I often call an *"Atmospheric Tantrum."* In such a world, the distinction between natural disaster and human error matters less than the speed with which we are able to react. The SHIELD was designed with this reality in mind. It does not attempt to determine the cause of a fire. Instead, it responds to the physics of heat, smoke, and environmental signals, allowing it to react immediately to any threat that approaches the perimeter of a protected home.

The system's strength lies in its independence. Unlike traditional emergency systems that rely on centralized command centers, distant fire departments, or fragile electrical grids, the SHIELD operates as a self-contained guardian. Because it is not dependent on outside control, it cannot be disabled by the failure of a power grid or compromised by digital interference. This design reflects a principle that every pilot understands well: redundancy. In aviation, no single system is allowed to be the only line of defense. When external systems fail, internal systems must remain fully operational. The same philosophy applies here. Even if the external world is experiencing chaos—whether through wildfire, power outages, or infrastructure breakdown—the internal systems protecting a home must continue to function with certainty. This independence grants something that modern society has largely forgotten: local sovereignty. A home protected by the SHIELD is no longer waiting for rescue; it is actively defending itself.

Central to this capability is the durability of the system's sensors and control units. These components form what might be described as the "brain" of the SHIELD, constantly monitoring the environment for sudden changes in temperature, wind direction, or airborne particulates. To ensure reliability in extreme conditions, these sensors are encased in materials similar to those used to protect aircraft flight recorders—the so-called "black boxes." Such materials are designed to survive extraordinary heat, pressure, and impact, ensuring that critical information remains intact even in catastrophic situations. In the case of the SHIELD, this protective design allows the system to continue analyzing environmental conditions even when flames approach dangerously close. The sensors do not panic or hesitate; they simply continue to observe, calculate, and respond.

When the system detects rising heat levels or approaching fire activity, it automatically adjusts water pressure and distribution across the property's defensive network. Instead of dispersing water evenly, the SHIELD concentrates its response on the areas where heat signatures are strongest. This targeted reaction maximizes efficiency while conserving resources, allowing the system to sustain its defensive posture for longer periods. In effect, the home becomes surrounded by an intelligent perimeter that continuously adapts to the shifting behavior of fire.

Yet even the most advanced system cannot sustain itself indefinitely without human attention. Technology can extend human capability, infrastructure can stabilize societies, and knowledge can guide progress, but none of these elements operate forever without care. Stability requires stewardship. Every bridge must be inspected, every aircraft must be serviced, and every system designed for protection must be maintained by those who understand its importance. For this reason, the systems described throughout this book—the SHIELD Project, the Subterranean Arterial Network (SAN), and the Silent Seeds—were never intended to function as isolated innovations. They are part of a broader framework designed to support long-term resilience and peace. Their effectiveness depends not only on engineering, but also on the individuals and communities willing to maintain them.

Those who accept this responsibility become more than operators of technology. They become guardians. Guardianship is not defined by authority or power but by commitment to preservation and foresight. Guardians recognize that the stability they maintain today will determine the safety and prosperity of generations that follow. Modern society often assumes that infrastructure, once built, will simply continue to function. Roads are expected to remain smooth, electrical systems are expected to deliver uninterrupted power, and food supplies are expected to appear reliably in markets. In truth, every one of these systems requires continuous oversight. When maintenance is neglected, small vulnerabilities gradually expand into major failures.

The SHIELD Project was designed to minimize such vulnerabilities, but its reliability still depends on careful monitoring. Guardians periodically examine wind-powered energy systems to ensure that the sentinels continue to generate and store sufficient power. Water reserves must be observed, sensor networks inspected, and automated response systems tested to confirm that they remain ready for sudden environmental changes. By maintaining these systems before crises arise, guardians shift society away from a culture of reaction and toward a culture of preparation.

Beneath the surface, the Subterranean Arterial Network requires a different form of vigilance. Operating largely out of sight, this underground system functions as a hidden circulatory network for essential resources such as water and energy. Because much of its complexity remains invisible to the public, guardians ensure that pipelines remain secure, resource flows remain uninterrupted, and structural integrity is preserved against geological shifts or natural wear.

Their work ensures that even when storms, fires, or other disruptions affect the surface world, the steady movement of life-sustaining resources continues below ground.

Above the surface, the Silent Seeds require stewardship of a different kind. Their maintenance is ecological rather than mechanical. Guardians monitor soil health, biodiversity, and crop resilience, ensuring that landscapes remain productive without exhausting the ecosystems that sustain them. Through thoughtful agricultural practices and environmental restoration, the Silent Seeds transform fragile environments into resilient foundations for human life.

The true strength of this integrated framework lies in the way these three elements reinforce one another. The SHIELD Project provides visible protection, standing as a network of sentinels that generate energy while defending homes from immediate threats. The SAN functions quietly beneath the surface, ensuring the continuous flow of essential resources regardless of surface disruptions. Meanwhile, the Silent Seeds restore the environmental systems that support agriculture, water cycles, and biodiversity. Each component supports the others. When balanced properly, they create a resilient structure capable of sustaining communities through both environmental and technological challenges.

Guardians must therefore understand how these systems interact. If environmental systems collapse, infrastructure eventually becomes vulnerable. If infrastructure fails, protection systems lose their ability to function effectively. And if protection systems collapse, the stability required for environmental restoration may vanish as well. Maintaining balance between these elements becomes one of the most important responsibilities entrusted to those who care for them.

Preserving these systems also requires foresight. The world will continue to evolve. Climate patterns will shift, technologies will advance, and new challenges will emerge. Guardians must remain adaptable, ensuring that the systems evolve without abandoning the principles that make them effective. For the SHIELD Project, this may mean integrating improved detection technologies or more efficient energy storage. For the SAN, it may involve expanding underground corridors or strengthening pipelines against future environmental pressures. For the Silent Seeds, it means protecting biodiversity and adapting agricultural systems to changing climates.

Equally important is the preservation of knowledge. Technical documentation, environmental records, and operational procedures must be carefully protected and passed forward. Without this knowledge, even the most sophisticated infrastructure can fall into neglect. Guardians therefore serve not only as caretakers of machines and ecosystems, but also as protectors of the information required to sustain them.

No guardian serves forever. One of the most vital responsibilities within any long-term system is preparing the next generation to continue the work. Education becomes central to this process. Young engineers, environmental scientists, agricultural specialists, and community leaders must learn how these systems function and why they exist. More importantly, they must adopt the mindset of stewardship. When societies begin to view infrastructure not as temporary convenience but as a long-term responsibility, their understanding of progress changes. Success is no longer measured solely by rapid growth but by the stability left behind for those who follow.

Peace itself emerges from this stability. It is often described simply as the absence of conflict, but genuine peace is something far more durable. It is created when societies possess reliable systems that protect life, distribute resources fairly, and preserve the natural environments that sustain them. The integrated framework presented in this book was designed with exactly that goal in mind. The SHIELD reduces vulnerability to immediate threats. The SAN protects the continuity of essential resources. The Silent Seeds restore and stabilize the ecological systems upon which human life depends.

Guardians ensure that these elements remain balanced and operational. Through their stewardship, communities become stronger, resources remain secure, and the likelihood of conflict driven by scarcity or instability diminishes. Peace, in this sense, does not appear suddenly. It is built patiently through responsible design, careful maintenance, and long-term vision.

Every generation inherits the work of those who came before. The roads we travel, the farms that feed us, and the technologies we depend upon all exist because someone chose to build systems intended to last beyond their own lifetime. The SHIELD Project, the SAN, and the Silent Seeds continue that tradition. They are not merely solutions for the present moment; they are frameworks intended to protect the future.

The guardians who maintain these systems stand quietly at the center of that responsibility. Their role is not simply to operate technology but to ensure that the foundations of stability remain intact long enough to benefit generations they may never meet. When they fulfill that responsibility, they create something rare in human history: a civilization that no longer spends its energy rebuilding after crisis, but instead invests its strength in strengthening the foundations of peace.

In that quiet and enduring work lies the true legacy of the guardians—a legacy not of conquest or domination, but of foresight, care, and balance. Through that balance, the vision of a stable and peaceful world moves steadily from possibility toward reality.

APPENDIX A

The Global Connectivity Map

Mapping the Integrated Network Between the SHIELD and the Subterranean Arterial Network (SAN)

The effectiveness of any large-scale infrastructure system depends not only on the strength of its individual components but on the precision with which those components are connected. The Global Connectivity Map illustrates the structural relationship between two central systems described throughout this book: the SHIELD Project and the Subterranean Arterial Network (SAN). While each of these systems can function independently, their full capability emerges when they operate as a coordinated network spanning regions, nations, and continents.

This map is not simply a geographic representation. It serves as a strategic visualization of how environmental protection systems on the surface interact with the hidden infrastructure that moves resources beneath the earth. Each point where a SHIELD sentinel connects to the SAN represents a node of stability—an engineered intersection where energy production, water distribution, environmental monitoring, and automated protection systems operate together.

Viewed in its entirety, the Global Connectivity Map reveals an architecture that extends far beyond individual communities. It outlines a planetary framework in which localized protection systems are supported by an underground resource network designed to maintain stability even when surface conditions become unpredictable. Through this integration, the visible guardians above ground and the hidden corridors below the earth function as a single, continuous system designed to protect life and sustain environmental balance.

Purpose of the Global Connectivity Map

The Global Connectivity Map fulfills several important roles within the integrated infrastructure model presented in this book.

First, it provides a clear structural overview of how SHIELD installations and SAN corridors interact across large geographic regions. By identifying every sentinel location and its corresponding underground access point, engineers and planners can visualize how energy generation, water resources, and environmental response systems are distributed throughout the landscape.

Second, the map allows for efficient monitoring of environmental conditions. Sensors located within SHIELD sentinels constantly collect data regarding temperature shifts, atmospheric conditions, water availability, and wildfire risk. When these sensors detect changes that require attention, the map allows technicians and monitoring centers to immediately identify which regions are connected and which systems may need to respond. This capability allows threats to be addressed quickly, often before they spread across wider areas.

Third, the map serves as a planning tool for long-term infrastructure expansion. As communities grow, environmental restoration zones expand, and climate patterns evolve, new SHIELD sentinels and SAN corridors must be installed strategically. The Global Connectivity Map ensures that expansion occurs in an organized and coordinated manner rather than through isolated development. Each new installation becomes part of a broader system designed to strengthen regional and global stability.

Perhaps most importantly, the map demonstrates that the system was designed to function as a unified structure. Rather than operating as scattered installations, each component contributes to a larger network of protection and sustainability.

Continental Integration Nodes

At the largest scale, the integrated system is organized through continental coordination hubs known as Primary Integration Nodes. These nodes act as major connection points where multiple SHIELD installations intersect with major SAN transport corridors.

Each Primary Integration Node serves as a regional anchor for infrastructure coordination. Within these nodes, energy generated by wind-powered sentinels can be directed into underground storage systems or distributed to nearby regions that require additional power. Water resources stored within subterranean reservoirs can be redirected to areas experiencing environmental stress, drought conditions, or wildfire risk.

In addition to energy and water management, these nodes collect environmental data from hundreds of SHIELD sentinels positioned throughout surrounding regions. This data is compiled and analyzed to provide a clearer understanding of atmospheric conditions, environmental stability, and infrastructure performance.

Because each continent contains its own network of integration nodes, regions maintain operational independence while remaining connected to the global system. This balance ensures that communities can respond quickly to local environmental conditions while still benefiting from the broader coordination of the planetary network.

Regional Sentinel Networks

Extending outward from each Primary Integration Node are clusters of Regional Sentinel Networks. These networks consist of multiple SHIELD windmill sentinels positioned strategically across diverse landscapes.

Sentinels are typically installed in areas where environmental monitoring and protection are most critical. These locations may include coastal wind corridors where renewable energy potential is strong, wildfire-prone mountain regions, agricultural zones that depend on stable water supplies, desert areas undergoing ecological restoration, and regions near populated communities that require enhanced protection.

Although each sentinel operates autonomously, its connection to the SAN allows it to participate in a much larger protective system. When one sentinel detects an environmental threat—such as rising heat signatures associated with wildfire conditions—neighboring installations receive automated alerts through the network. This allows multiple systems to respond simultaneously.

For example, several nearby sentinels may activate cooling or saturation systems while underground water reservoirs increase pressure within the SAN's distribution lines. Through this coordinated response, environmental threats can be addressed quickly before they expand into large-scale disasters.

The SAN Connection Interface

At the foundation of each SHIELD sentinel lies a vertical interface that connects the surface system to the Subterranean Arterial Network below. This interface acts as the critical bridge between surface monitoring technology and underground infrastructure.

Each interface typically contains three integrated channels that support the operation of both systems.

The first channel is the energy conduit. Wind energy captured by the sentinel's turbine is transmitted downward through this conduit into the SAN's energy grid. From there, the energy may power underground pumping stations, sensor networks, monitoring hubs, and regional storage facilities.

The second channel functions as a water and resource riser. Pressurized water stored within subterranean reservoirs travels upward through this pathway to supply the SHIELD's automated sprinkler systems, cooling mechanisms, and emergency response tools. This vertical flow ensures that protective resources are always available at the surface without relying on external water delivery systems.

The third channel serves as a data and sensor transmission link. Environmental readings collected by the sentinel—including atmospheric data, soil moisture levels, and thermal measurements—are transmitted into the SAN's communication network. These data streams allow monitoring centers and nearby installations to remain informed about environmental conditions across wide regions.

Together, these three channels transform each sentinel into both a protective guardian on the surface and an access point to the underground infrastructure supporting it.

Global Monitoring and Research Centers

Although the integrated network is designed to operate autonomously, several Global Monitoring and Research Centers provide long-term observation and analysis.

These centers do not function as centralized command authorities. Instead, they operate as scientific and technical institutions dedicated to studying environmental patterns, evaluating system performance, and developing improvements to infrastructure design.

Researchers within these centers analyze global sensor data to identify emerging climate patterns, assess the resilience of energy systems, and recommend upgrades to detection technologies. They also serve as hubs for international collaboration, allowing engineers and environmental scientists from different regions to share knowledge and coordinate infrastructure development.

By maintaining a global perspective on environmental and technological trends, these centers ensure that the integrated network continues evolving alongside the challenges it was designed to address.

Visual Structure of the Connectivity Map

Within the full design documentation of the system, the Global Connectivity Map is presented as a layered visual blueprint.

The surface layer identifies the geographic locations of SHIELD windmill sentinels installed across continents. The subsurface layer illustrates the SAN's underground corridors, reservoirs, and transport pathways. Additional layers highlight continental integration nodes, environmental restoration zones supported by Silent Seeds, and the locations of monitoring and research centers.

This layered visualization allows engineers, planners, and system guardians to observe how the surface and subterranean systems interact within a single framework. It also helps reveal the scale of the infrastructure and the degree of coordination required to maintain it.

A Network Designed for Cooperation

Ultimately, the Global Connectivity Map represents more than a technical diagram. It reflects a guiding principle behind the entire system: stability grows stronger when systems are designed to cooperate rather than compete.

By linking energy generation, resource distribution, environmental monitoring, and protective response mechanisms into a single network, the SHIELD Project and the Subterranean Arterial Network create an infrastructure capable of supporting both human societies and natural ecosystems.

Each sentinel installed and each SAN corridor constructed strengthens the architecture of stability. As new connections are added, the network expands into a cooperative framework that transcends national boundaries and regional limitations.

Over time, this expanding infrastructure forms something remarkable: a planetary system designed not for conflict, but for protection, restoration, and continuity.

The Global Connectivity Map simply reveals what has always been the deeper intention behind this vision—a world in which the infrastructure of peace extends across the entire planet, quietly sustaining the balance between humanity and the environment.

APPENDIX B

The Fiji History Archives
Ancient Environmental Knowledge and the Foundations of Modern Infrastructure

The technologies and infrastructure systems described throughout this book may appear entirely modern at first glance. The Subterranean Arterial Network (SAN), the SHIELD Project, and the environmental restoration efforts represented by the Silent Seeds rely on advanced engineering, modern materials, and sophisticated monitoring technologies. Yet many of the principles that guide these systems are not new. They echo ideas that have existed for centuries in communities that learned to survive by carefully observing and cooperating with the natural world.

The Fiji History Archives provide historical context for this connection between past knowledge and modern innovation. This collection of historical records, field notes, reconstructed diagrams, and geological studies documents how ancient communities within the islands of Fiji developed systems for managing water, land, and environmental stability. While these early systems were created without modern machinery, they reveal an impressive understanding of natural resource management and landscape engineering.

Rather than viewing these historical methods as primitive, the archives show them to be examples of highly adaptive environmental intelligence. Communities that lived within the volcanic landscapes of Fiji learned to design infrastructure that worked with the terrain rather than against it. Their survival depended on understanding how water moved through rock formations, how soil responded to seasonal cycles, and how ecosystems maintained their own balance.

In many ways, the modern SAN reflects this same philosophy. While the materials and technologies have evolved, the fundamental idea remains similar: life-sustaining resources should move through protected pathways that cooperate with natural geological structures.

By examining the historical practices documented in the Fiji History Archives, we gain insight into how ancient environmental knowledge continues to influence the infrastructure systems designed for the future.

The Volcanic Sluice Maps of Taveuni

Among the most significant discoveries preserved within the Fiji History Archives is a series of reconstructed diagrams known as the Volcanic Sluice Maps. These diagrams focus primarily on the island of Taveuni, a region shaped by ancient volcanic activity and characterized by extensive basalt rock formations.

Early Fijian communities recognized that these volcanic landscapes offered unique opportunities for controlling the movement of water. Rather than attempting to redirect rivers or construct large surface canals, they utilized the natural fractures and porous structures within basalt rock to guide water beneath the surface. Through careful carving and shaping of underground channels, they created controlled pathways that allowed water to travel across the terrain while remaining protected from evaporation and environmental disruption.

These channels functioned as early forms of subterranean water infrastructure. Rainwater collected in higher elevations would gradually flow through underground conduits and emerge in lower agricultural areas, providing a steady and reliable source of irrigation. During periods of heavy rainfall, additional channels allowed excess water to disperse safely, preventing flooding and protecting farmland.

The volcanic rock itself played an important role in this system. Basalt formations acted as natural protective barriers, shielding water from contamination while maintaining cool temperatures within the channels. Because the water remained underground, it was far less vulnerable to the heat of the tropical sun or the unpredictable conditions of the surface environment.

The Volcanic Sluice Maps illustrate these networks in remarkable detail. The diagrams show branching channel systems extending across the landscape, connecting elevated reservoirs with agricultural terraces and village settlements. Each pathway appears to have been carefully designed to follow the natural contours of the terrain, allowing gravity to assist the flow of water without requiring mechanical assistance.

These systems reveal a deep understanding of hydrology and geological structure. Instead of forcing water to behave unnaturally, ancient engineers worked with the existing characteristics of the landscape.

This philosophy strongly influenced the conceptual design of the Subterranean Arterial Network. While the SAN incorporates modern engineering techniques and advanced monitoring systems, it follows a similar principle: resources move most efficiently when infrastructure works in harmony with natural geological conditions.

The "Luvu" Engineering Principles

In addition to the Volcanic Sluice Maps, the Fiji History Archives preserve descriptions of a set of traditional environmental design practices referred to collectively as the Luvu Principles. These principles represent a collection of ideas and techniques that guided early infrastructure development throughout parts of the Fijian islands.

Although these concepts were developed centuries ago, they reflect a surprisingly sophisticated understanding of sustainable engineering. Many of their underlying ideas align closely with modern approaches to environmental resilience and resource efficiency.

One of the central Luvu principles involves the use of natural thermal regulation. Ancient builders observed that underground environments maintained more stable temperatures than surface conditions. By placing water channels and storage areas beneath the earth, they protected valuable resources from extreme heat and evaporation. The surrounding soil and rock acted as natural insulation, preserving cooler temperatures without the need for artificial cooling systems.

A second principle emphasized gravity-guided resource movement. Instead of relying on complex mechanical devices, early engineers designed water pathways that followed natural elevation gradients. This allowed water to move slowly and consistently across the terrain using gravity as the primary driving force. By carefully mapping the landscape, they ensured that water reached agricultural fields and settlements with minimal energy expenditure.

Another key concept involved layered geological protection. Basalt rock and other volcanic formations were used strategically to reinforce underground structures. These natural materials provided strength and durability while protecting channels from collapse, erosion, and environmental damage. Even without modern reinforcement materials, these geological layers created remarkably resilient systems.

The Luvu Principles also emphasized distributed resource storage. Rather than depending on a single reservoir, water was stored in multiple locations throughout the landscape. This decentralized approach ensured that if one source became damaged or depleted, others could continue supplying nearby communities.

Perhaps the most important Luvu principle was the idea of environmental harmony. Infrastructure was designed to complement the surrounding ecosystem rather than dominate it. Agricultural zones were located where soil conditions were strongest, forests were preserved to protect watersheds, and settlements developed in areas that balanced human needs with environmental stability.

These ideas remain deeply relevant today. The SAN, the SHIELD Project, and the Silent Seeds all reflect the same fundamental belief: infrastructure must work with the natural world rather than attempting to overpower it.

Bridging Ancient Knowledge and Modern Engineering

The purpose of including the Fiji History Archives in this book is not merely historical curiosity. It serves as a reminder that innovation often emerges from rediscovery.

Ancient societies developed solutions to environmental challenges using the tools and knowledge available to them. Through careful observation of natural systems, they learned how landscapes functioned and how human infrastructure could coexist with those systems.

Modern engineering possesses capabilities that those societies could never have imagined. Advanced materials, renewable energy technologies, digital sensors, and automated monitoring systems allow us to build infrastructure on an unprecedented scale. Yet the wisdom gained from earlier generations remains valuable.

When these two sources of knowledge are combined, a powerful synergy emerges.

The Subterranean Arterial Network represents this synthesis. It merges the ancient logic of underground water channels with modern resource transport systems. It incorporates natural thermal regulation and geological protection while adding digital monitoring and renewable energy integration.

In this way, the SAN becomes more than a technological system. It becomes a bridge between historical environmental intelligence and the engineering capabilities of the modern world.

Preserving the Legacy of Environmental Knowledge

The Fiji History Archives also highlight a broader lesson: knowledge must be preserved if it is to guide future generations.

Just as modern guardians maintain the infrastructure described throughout this book, historians and researchers preserve the insights of past civilizations. Every map, diagram, and field note contained within the archives represents a fragment of understanding about how human societies once interacted with the environment.

When these fragments are studied carefully, they reveal patterns that remain relevant even today.

They remind us that humanity has long sought ways to live sustainably within the natural world. The technologies may change, but the underlying goals remain the same: stability, resilience, and balance between human communities and the ecosystems that sustain them.

In this sense, the volcanic channels of ancient Fiji and the underground corridors of the Subterranean Arterial Network share a common purpose—to ensure the reliable movement of life-sustaining resources while preserving the integrity of the environment.

APPENDIX C

The Shatterproof Oath

A Commitment to Stewardship and Peace

Throughout this book, we have explored systems designed to protect communities, preserve environmental stability, and support the long-term resilience of civilization. The SHIELD Project provides visible protection from environmental threats. The Subterranean Arterial Network safeguards the continuous movement of vital resources. The Silent Seeds restore the ecological foundations upon which human life depends.

Yet no system—no matter how carefully designed—can sustain itself indefinitely without human responsibility.

Technology can assist humanity, but it cannot replace the values that guide human action. Infrastructure can strengthen societies, but it cannot guarantee that those societies will maintain the systems they inherit.

For this reason, the concept of guardianship stands at the heart of the framework presented in this book.

The Shatterproof Oath represents a symbolic expression of that guardianship. It is not a legal agreement or a formal contract of authority. Instead, it is a declaration of intention—a reminder that the systems created to protect the future must also be supported by individuals and communities who understand their responsibility to maintain them.

When spoken or read aloud, the oath serves as a reflection of the principles that guide the guardians of stability: stewardship, cooperation, knowledge, and respect for the natural world.

The Shatterproof Oath

We stand as guardians of balance.

We recognize that peace is not an accident of history, but the result of foresight, preparation, and care.

We commit ourselves to protecting the systems that sustain life the wind that powers our sentinels,the water that flows beneath our feet, and the seeds that restore the living earth.

We will maintain the SHIELD that protects our communities.

We will safeguard the hidden corridors of the Subterranean Arterial Network, ensuring that the resources of life continue to move safely through the world.

We will nurture the Silent Seeds that rebuild the soil, restore ecosystems, and strengthen the natural foundations of civilization.

We reject destruction as a path to security.

Instead, we choose knowledge, cooperation, and stewardship.

We understand that stability must be preserved through responsibility, and that the peace we inherit today must be strengthened for those who will follow.

We pledge to protect the balance between humanity and the environment.

We pledge to preserve the knowledge that guides these systems.

We pledge to pass this responsibility forward to future generations.

So that the systems we maintain remain strong.

So that the land we restore remains alive.

So that the peace we protect remains unbroken.

Together, we commit ourselves to a future where stability is not fragile, where protection does not rely on destruction, and where the foundations of civilization are built to endure.

This is our promise.

This is our responsibility.

This is the Shatterproof Oath.

CONCLUSION

THE HARVEST OF SAFETY

Throughout history, human civilization has been shaped by the systems it builds to protect itself. Cities have risen behind walls, nations have defended borders with armies, and entire technological eras have been defined by the tools designed to confront threats. Protection, for most of human history, has been associated with force—the ability to repel danger through strength or confrontation.

Yet the challenges facing the modern world reveal the limitations of that model. Many of the greatest threats confronting humanity today are not adversaries that can be defeated through traditional conflict. Wildfires, climate instability, water scarcity, ecological degradation, and fragile infrastructure systems are not enemies that retreat when confronted with weapons. They are complex environmental and structural realities that require a fundamentally different response.

This book has explored a new way of thinking about protection—one rooted not in destruction, but in stability.

The SHIELD Project, the Subterranean Arterial Network, and the Silent Seeds together form what can be described as an **architecture of peace**. Each system addresses a different dimension of resilience. The SHIELD provides visible protection from environmental threats, standing across landscapes as a network of sentinels capable of detecting danger and responding before catastrophe can spread. The Subterranean Arterial Network operates beneath the surface as the hidden circulatory system of civilization, ensuring that energy, water, and vital resources continue to flow even when surface conditions become unstable. Meanwhile, the Silent Seeds restore the ecological foundations upon which human life ultimately depends, strengthening soil systems, rebuilding biodiversity, and stabilizing agricultural environments.

Individually, each of these systems offers a form of protection. Together, they form a comprehensive framework designed to support long-term resilience.

What makes this framework truly significant, however, is not only the technology involved. It is the philosophy behind it. The systems described in these pages are built upon a simple but powerful idea: the most effective forms of protection are those that prevent crises from occurring in the first place.

Prevention requires foresight. It requires understanding how environmental systems behave, how infrastructure can fail, and how small disturbances can grow into large-scale disasters if left unchecked. By designing systems that monitor environmental conditions continuously, distribute resources efficiently, and restore ecological balance over time, societies can reduce the likelihood that many crises will escalate beyond control.

In this sense, the architecture of peace is not passive. It is an active commitment to stability.

The SHIELD windmill sentinels demonstrate this concept in a visible way. Rising above landscapes across hills, valleys, and coastlines, they harvest wind energy while simultaneously observing the environment around them. Their sensors monitor atmospheric conditions, temperature changes, and early warning signals that might indicate environmental danger. When abnormal patterns appear, the system can respond quickly, deploying cooling mechanisms or resource support systems before a small disturbance evolves into a large-scale disaster.

Below the surface, the Subterranean Arterial Network reinforces this stability by ensuring that life-sustaining resources remain secure and accessible. Water reservoirs protected by the earth's natural insulation remain cool even under extreme heat conditions. Energy captured by windmill sentinels can be distributed through underground corridors that remain insulated from storms, fires, and surface disruptions. These subterranean pathways allow communities to maintain essential services even when visible infrastructure becomes compromised.

At the same time, the Silent Seeds represent the long-term ecological foundation that sustains the entire system. By restoring degraded landscapes, strengthening soil ecosystems, and encouraging biodiversity, they rebuild the natural resilience of the environment itself. Over time, this restoration reduces the likelihood of catastrophic fires, improves water retention in the soil, and stabilizes the agricultural systems that support human populations.

Together, these systems demonstrate how infrastructure can evolve from a reactive model into a proactive one.

Instead of waiting for disasters to occur and then mobilizing emergency responses, societies can design systems that anticipate risk and respond early. The result is a shift in how security itself is understood. Protection becomes less about confrontation and more about preparation.

Yet even the most advanced infrastructure cannot sustain itself without human stewardship. Technology may provide tools, but responsibility remains a human choice.

The concept of guardianship therefore stands at the center of the framework presented in this book. Guardians are not defined by authority or power. They are defined by commitment—by the willingness to maintain the systems that protect both human communities and the natural world.

Engineers who maintain energy systems, environmental scientists who monitor ecosystems, agricultural stewards who protect soil health, and local communities who support resilient infrastructure all become part of this guardianship. Each plays a role in preserving the stability that future generations will depend upon.

The Shatterproof Oath described in the appendix reflects this responsibility. It reminds us that the infrastructure of peace must be supported by the intention of the people who maintain it. Systems alone cannot guarantee stability; they must be guided by individuals who recognize their role in protecting both the present and the future.

This understanding transforms the way societies think about progress.

For much of modern history, progress has often been measured by speed—faster technology, faster production, faster expansion. While these achievements have produced remarkable innovations, they have sometimes overlooked the importance of durability. Systems designed for rapid growth are not always designed for long-term stability.

The architecture of peace offers a different perspective. Instead of asking only how quickly a society can advance, it asks how long its foundations can endure. Infrastructure becomes an investment not just in immediate development but in generational continuity.

When energy systems are renewable and decentralized, communities become less vulnerable to disruption. When water resources are protected underground, they remain available even during extreme environmental conditions. When ecosystems are restored rather than exhausted, the natural world becomes an ally rather than a liability.

These changes do not eliminate all challenges, but they significantly strengthen humanity's ability to navigate them.

The vision described throughout this book is ambitious, yet its foundation is practical. None of the individual components— renewable energy generation, underground infrastructure, ecological restoration, or distributed monitoring systems—are beyond the reach of modern engineering. The real challenge lies not in technological feasibility but in collective commitment.

Building the architecture of peace requires cooperation across communities, regions, and nations. Environmental systems do not recognize political boundaries, and the stability of one region often depends on the health of ecosystems elsewhere. By designing infrastructure that supports both local independence and global cooperation, societies can strengthen resilience on a planetary scale.

The Global Connectivity Map described in Appendix A illustrates how such coordination might unfold. As networks of SHIELD sentinels and SAN corridors expand across continents, they form a distributed system capable of supporting both environmental stability and resource security. Each connection point represents not only an engineering achievement but also a symbol of collaboration between communities working toward shared resilience.

Over time, these networks could transform how humanity interacts with the planet itself.

Instead of viewing the environment primarily as a resource to be extracted, societies may increasingly see it as a system to be stabilized and protected. Infrastructure will not simply support economic activity but will also reinforce ecological balance.

This shift in perspective may ultimately become one of the most important developments of the coming century.

The future will undoubtedly present new challenges—technological, environmental, and social. Yet the solutions to those challenges will depend largely on the systems that humanity chooses to build today.

If the infrastructure of the future is designed with foresight, resilience, and cooperation in mind, then the foundations of civilization will grow stronger with each generation. Communities will inherit systems that protect them not through force, but through stability.

That is the promise of the architecture described in this book.

A future where wind-powered sentinels watch over the landscape.

Where life-sustaining resources move safely beneath the earth through resilient underground corridors.

Where restored ecosystems strengthen the soil, water, and biodiversity upon which human life depends.

And where the guardians of these systems understand that their responsibility extends beyond their own lifetimes.

In such a world, peace is not merely the absence of conflict. It becomes the result of thoughtful design, careful stewardship, and the shared recognition that the stability of civilization depends on protecting both people and the planet they inhabit.

The systems described here are only the beginning of that vision.

But every structure begins with a foundation, and every lasting future begins with the decision to build wisely.

AUTHOR BIOGRAPHY

Mahendra Jagir is a visionary engineer, environmental strategist, and systems thinker dedicated to guiding humanity toward a future defined not by conflict and instability, but by structural sustainability and lasting peace. Throughout his life, Jagir has pursued a singular question: how can human ingenuity be used not only to build cities and machines, but to design a civilization capable of protecting itself while living in balance with the natural world.

Drawing upon his deep connection to the history, geography, and cultural traditions of Fiji, Jagir integrates ancient ecological wisdom with modern engineering principles. The island societies of the Pacific developed remarkable methods of environmental stewardship long before the modern era, and Jagir has long believed that these forgotten insights can help inform solutions to today's global challenges. By studying these traditions alongside contemporary technological innovation, he has developed a philosophy that merges heritage with forward-thinking design.

Central to this philosophy is Jagir's pioneering work on two ambitious systems: **the SHIELD Project** and the **Subterranean Arterial Network (SAN)**. Together, these concepts form the technological foundation for a new model of resilience. The SHIELD Project introduces a defensive infrastructure capable of protecting communities from environmental threats such as wildfire and extreme climate conditions, while the SAN envisions a global underground network that ensures the reliable distribution of water, energy, and essential resources. Through these integrated systems, Jagir proposes that peace and security are not merely political ideals but engineering challenges that can be solved through thoughtful design.

Jagir's work reflects decades of study, observation, and reflection on the vulnerabilities of modern civilization. As climate change, environmental degradation, and resource instability continue to shape the global landscape, he argues that humanity must move beyond reactive crisis management and instead build systems that anticipate and neutralize these threats before they escalate. In Jagir's vision, infrastructure itself becomes a guardian—quietly working beneath the surface to maintain stability and protect future generations.

The Shield and the Seed represents the culmination of this vision. More than a technical proposal, the book serves as both a practical blueprint for climate resilience and a deeply personal legacy. Written with his grandchildren in mind, the work reflects Jagir's hope that the next generation will inherit a world strengthened by foresight rather than weakened by neglect.

Through his writing, Jagir challenges readers to reconsider the priorities that have shaped modern development. For too long, humanity has invested enormous resources preparing for destruction—building weapons, fortifying borders, and reacting to disasters only after they occur. Jagir calls for a different path: one where human creativity is directed toward building systems that protect life, restore ecosystems, and stabilize the environment.

At the heart of his philosophy lies a powerful idea—that peace is not simply the absence of conflict, but the presence of strong, intelligent systems that allow societies to thrive. By combining environmental restoration, sustainable infrastructure, and advanced engineering, Jagir believes humanity can create what he calls **"Shatterproof Peace"**—a future where the foundations of civilization are resilient enough to withstand the uncertainties of a changing world.

The Guardian Declaration

At the closing of this work, we pause to acknowledge a truth that has guided every page of this book: the Earth is not ours to possess. It is entrusted to us through time—shaped by those who came before us and sustained only through the care of those who will follow.

Every generation inherits the world in two forms: as a gift and as a responsibility. The forests, waters, soils, and skies that sustain life are not merely resources to be consumed, but foundations that must be preserved with wisdom and foresight. The choices we make today shape the stability of tomorrow.

The systems described throughout this book—the SHIELD, the Subterranean Arterial Network, and the Silent Seeds—represent more than technological ideas. They form a philosophy of protection, resilience, and restoration designed to safeguard both humanity and the natural world.

The **SHIELD** stands as the visible guardian. It represents preparedness, vigilance, and the understanding that safety is not accidental but the result of careful design. Through the intelligent use of water, energy, and environmental monitoring, the SHIELD transforms infrastructure into a peaceful instrument of protection.

The **Subterranean Arterial Network (SAN)** forms the hidden lifeline beneath the surface of the Earth. Just as arteries sustain the human body, this underground network carries the essential elements of civilization—water, energy, and communication—through protected pathways that remain stable even when the surface world becomes uncertain.

The **Silent Seeds** embody the promise of renewal. They remind us that lasting peace cannot exist without a healthy planet. By restoring forests, strengthening soils, and rebuilding ecosystems, these seeds symbolize the quiet yet powerful process through which nature heals itself when given the chance.

Together, these three pillars form a unified vision for the future—one in which protection, infrastructure, and environmental restoration work in harmony.

But systems alone cannot guarantee the future.

Every generation must choose whether it will act as a consumer of the Earth or as a guardian of it.

The Shatterproof Oath is a declaration that the responsibility for this future is accepted with clarity and intention.

To sign this oath is to recognize that true security is built not through conflict, but through preparation, stewardship, and cooperation with the natural systems that sustain life.

It is a promise to preserve the balance between technology and nature.

It is a commitment to maintain the systems that protect communities and restore the land.

And above all, it is a pledge to leave the world stronger, healthier, and more stable than we found it.

Through this declaration, the first guardians affirm their role in carrying this responsibility forward.

Their signatures represent not only their names, but their commitment to the generations that will one day inherit the Earth.

The First Guardians

1. ———————————————

Dustin

2. ———————————————

Matthew

3. ———————————————

Georgia

4. ———————————————

Jolie

Witnessed and Dedicated By

Mahendra Jagir

"The surface is ours to enjoy,
but the future is ours to protect."

The Guardian Declaration

Every journey eventually reaches a quiet moment where the engines slow, the horizon steadies, and we reflect on the path that brought us there. As I write these final words, I think not of the miles I have traveled across oceans and continents, nor of the systems and ideas described throughout these pages, but of the simple hope that inspired them all: the desire to leave the world safer, stronger, and more balanced than I found it.

Throughout my life I have moved between two worlds—the open sky above and the silent strength of the earth below. From the cockpit of an aircraft, I learned how fragile the surface of our planet can appear when viewed from great heights. Yet beneath that fragile surface lies immense resilience, a quiet stability that has supported life for countless generations. The systems described in this book—the SHIELD, the Subterranean Arterial Network, and the Silent Seeds—are reflections of that lesson. They are attempts to design our future with the same wisdom that nature has practiced for millennia: protect what sustains life, restore what has been damaged, and prepare carefully for the uncertainties ahead.

But no design, no technology, and no blueprint can replace the most important element of all: human responsibility. The future of this planet will not be determined by machines alone, but by the choices of the people who inherit it. Every generation must decide whether it will merely consume the gifts of the Earth, or whether it will stand as a steward of them.

To my grandchildren—Dustin, Matthew, Georgia, and Jolie—these pages are my message across time. I hope that when you walk beneath the turning blades of a windmill, or stand beneath the shade of a growing forest, you will understand that these things were built not out of fear of the future, but out of faith in it. Faith that the world can become more peaceful, more balanced, and more resilient when people choose cooperation over conflict and stewardship over neglect.

If these ideas continue beyond these pages—if they inspire even a few people to build, restore, and protect—then the work has been worthwhile.

The systems are prepared

The seeds are planted.

The guardians now stand watch.

The future, from this moment forward, belongs to you.

www.ingramcontent.com/pod-product-compliance
Lightning Source LLC
Chambersburg PA
CBHW050012040726
47599CB00014B/1345